CW01021520

# The Badger Book

by Jo Byrne

Series editor Jane Russ

GRAFFEG

# Dedication

Dedicated to Wounded Badger Patrollers and Pixies across all of the cull zones. Thank you for your passion and your determination. Thank you for every single badger saved.

# Contents

# Introduction

I have an unbelievably soft spot for this noble yet elusive mammal. Their shy nature, bumbling sort of gait and iconic colouring. Unfortunately, like many of us, it's mostly online videos I can watch of them larking about as cubs outside their setts or snuffling around in people's backyards enjoying snacks and treats left out for them. They seem gentle and family-oriented, funny yet fierce, but are, unfortunately, demonised by the government and dairy industry over bovine tuberculosis (bTB). In fact, in one form or another, badgers have been brutalised by hunters and baiters for centuries.

My first ever protest. My first ever demonstration. The first time I felt enraged enough to risk the wrath of the authorities and physically stand against a government decision. I was protesting the badger cull outside Westminster – nervously chanting along with the others but enthralled and excited about the turnout. Surely the government, DEFRA and Natural England would hear the impassioned calls for a halt to the bloody killing? Surely they would consider the science that showed that whilst badgers *could* transmit bovine tuberculosis (bTB) to cattle, it didn't necessarily mean there *were* transmitting it.

The protests fell on deaf ears.

By the start of the cull season the following year, I'd given up shouting outside Westminster and marching to the DEFRA offices proffering petitions and pleas. By then, I'd donned my walking boots and hi-vis and joined Wounded Badger Patrols in the cull zones wherever and whenever I could –

civil resistance groups made up of ordinary, everyday people like myself determined to try save the badgers. While the patrollers are out there trudging around in the middle of the night, the shooters are not allowed to free-shoot and must, instead, rely on killing cage-trapped badgers for their bounty fees.

There were 64,000 badgers on DEFRA's hit list for 2020 – in 54 cull zones, across 16 counties. But the resistance goes on and I will continue to resist until the slaughter ends.

In this book we'll cover the issue of bTB and look at the science – offered by both sides – as to the reasons for the cull but we'll also meet some of the people determined to resist them. After all the seriousness of badger bloodshed, we'll take a look at badgers in myth and legend as well as art and literature.

First, however, we'll look at the physiology of the badger, how they've evolved over the millennia, how they live together (but not actually *live* together!) and some of the many, many fascinating facts about this enigmatic mammal.

Did you know, badgers can 'choose' when to fall pregnant? Furthermore, if undisturbed, badgers will occupy the same setts for hundreds and hundreds of years! Badgers often prefer to live with the clan they were born into for the rest of their lives and they even have a 'favourite' earthworm!

Hopefully – despite the culls and the baiters and the hunters – we can trust the words of Badger from Kenneth Grahame's classic *Wind in the Willows*: 'We are an enduring lot, and we may move out for a time, but we wait, and are patient, and back we come. And so it will ever be.'

# Badger Physiology

# Badger Physiology

Badgers are the land-tanks of the Mustelidae family. Whilst most of the group are thin and slinky and fast – like weasels, ferrets, otters and pine martins – the badger is short and stout with powerful legs. The European, or Eurasian, Badger (*Meles meles*) grows the heaviest of the badger species, as much as 20kg or so when fattened up in autumn – as well as the biggest – up to 30cm tall and up to 90cm long. Effectively, the size of a big, heavy dog!

Badgers are the largest carnivorous mammals in the UK.

The European badger is found across Europe and into parts of Asia. Other than some of the islands, they are found throughout most of the UK. Some areas, like south west England, are more densely populated than others. They're less common on the flatlands of East Anglia because they're, wisely, very particular about soil conditions. Clay soil risks flooding and anything too chalky poses a risk of setts collapsing.

## In the Beginning...

It is thought that badgers began to evolve separately from their long-tailed, climbing cousins within the Mustelid family around 20 million years ago. The earliest fossil remains dated back 2 million years have been found in France. Bones unearthed in West Sussex suggest that brock has been with us for the last 750,000 years.

Subsequent glacial periods forced the badgers out of northern Europe, sending them to regions we now know as Italy and Spain. Around 10,000 years ago, melting glaciers

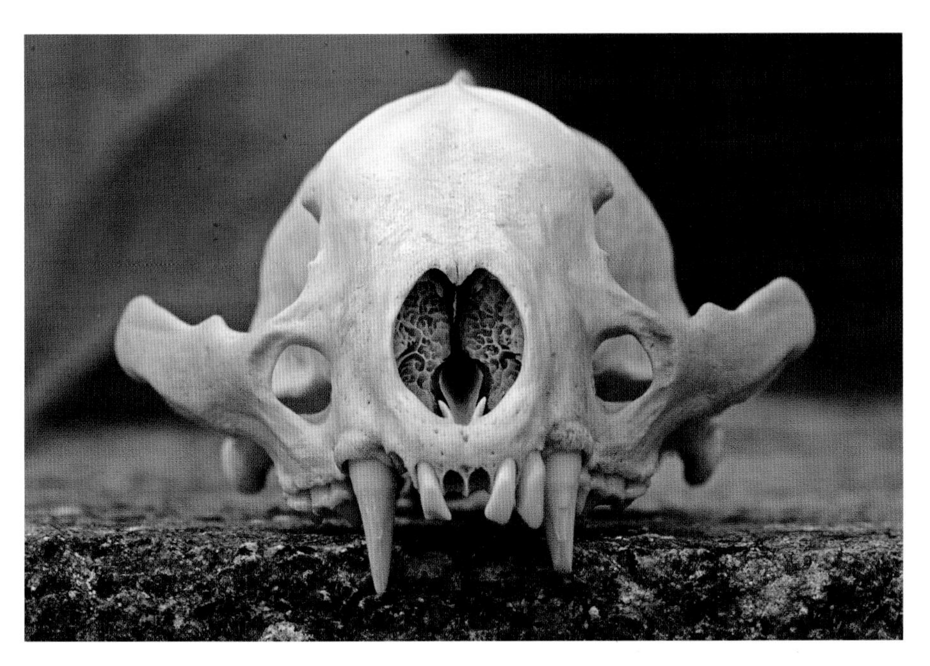

gave the badger the opportunity to rootle and bumble their way back to Britain, where they've been found ever since.

A male badger is known as a boar; a female badger is called a sow and young are known as cubs.

Like other mustelids, badgers have strong anal scent glands – two inside the anus and one deep subcaudal gland just above the anus. These glands produce strong, musky, smelly secretions which the badgers use to recognise each other and their territories. Each clan has a unique scent. The badgers will squat briefly against each other and leave their scent on each other as well as around their sett, their latrines and along their well-worn pathways. The dominant boar in each clan, or colony, of badgers will scent more than the others. Badgers have even been known to scent-mark humans they've become accustomed to!

The name 'badger' is thought to derive, simply, from the word 'badge' in reference to their distinct facial markings being like a badge, emblem or decoration. Other references indicate the word might come from the old French words *bêcheur*, meaning 'digger', or *blaireau,* meaning 'corn hoarder'.

A common nickname for our noble beast is 'brock' – derived from Old English (brocc), Middle English (broc) – and has lent itself to the naming of various towns and villages across the UK, including Brockenhurst in Hampshire, and Brock (next to the River Brock) in Lancashire.

## You Lookin' at Me?

With its white face and bandit mask, the badger's distinct facial pattern is truly unmistakable. The rest of the body is covered in a dense white underfur overlaid by coarse guard hairs which are largely white, save for a black band near the tip, giving the badger an overall grey, almost salt 'n' pepper colouring. The legs and underbelly are usually black. Badgers will fluff up their fur if they get a fright or are alarmed.

There are also ginger (erythristic) badgers that have light brown or distinct red fur (see opposite). White badgers (either leucistic or albino), in which their black pigment is replaced by white, are rare but not impossible to find.

Overall, it is a stocky, low-slung and powerfully built mammal. With a tube-like shape, thick neck, strong shoulders and front legs sporting long, non-retractable claws, the badger's design is perfect for digging and tunnelling.

Badgers have small eyes, which would be expected of an animal enjoying a good portion of life underground, but their sight is adequate. They have small ears as well but with good hearing. It's a badger's sense of smell, however, that makes theirs a world of scent rather than sound or vision. The wet, flexible nose is around 800 times better than ours at sniffing out information and is one of the largest noses in the mustelid family. A generous nasal cavity filled with small, heavily pleated bones makes for a large surface area for picking up smells. Sensitive whiskers are found around the nose and eyes.

The fearsome bite of the badger is assisted by its sagittal crest – a prominent ridge of bone running lengthways across the top of the skull. The thick muscles that develop along this ridge attach to powerful muscles on either side of the lower jaw. The majority of a badger's diet has now developed into softer foods, such as earthworms and fruits, but the powerful bite of the badger is still formidable in fighting and self-defence.

Generally, boars and sows are similar in size, although boars are marginally thicker in the head and neck and sows might be a smidge sleeker.

Whilst a badger's gait is usually a lumbering meander on their short, stumpy legs, they can happily trot over to their next feeding area or even canter if worried about a nearby danger. If needed, a badger can break into a quite speedy 25-30 kmph run for a short burst. They are also surprisingly proficient swimmers and even good climbers, particularly if there's food involved.

## Social vs Sociable

It turns out that, whilst badgers live together, they don't particularly look out for each other. And they don't look after or nurse each other's young either. Indeed, non-dominant sows risk their litters being wiped out by dominant sows. Nor do they share food. It seems that even though badgers are social – sharing a living space and territory – they're far from sociable and live alongside each other rather than with each other.

In the event of danger, one badger will not warn its nearby companion but will, instead, simply run off! And they certainly won't band together in a fight. Badgers prefer to forage on their own, perhaps sharing an area of plentiful worms but no more than that and will, largely, ignore each other should paths cross on their nightly forays.

Other than mutual grooming when leaving the sett in the evenings, it would seem the only other benefit badgers have to sharing a living space is for warmth in winter. Cozying up together in the main sett helps conserve vital energy and fat supplies to see them through the long cold days and nights. As soon as it warms up, badgers will split up and sleep separately!

They do have a distinct 'pecking order', with the dominant boar and sow retaining the primary breeding rights.

If frightened or to indicate submissiveness, a badger will make itself as small as possible and hide its face mask. At times of aggression, badgers will attempt to make themselves appear larger and ripple their fur at an opponent. They will crouch low and put their head down when ready to charge. If cornered, a badger will roll up to protect its head from attack.

Traditionally, there are even numbers of male cubs and female cubs born in each clan but due to fighting and run-ins with road traffic, males tend to die earlier than females, or simply evictions by dominant sows, meaning that most clans have a higher proportion of sows to boars.

Life expectancy for badgers varies, with a predictably shorter lifespan of around 6 to 10 years for wild badgers but ages well into the teens for those in captivity.

## Family Planning

In Britain, females usually ovulate between January and March as well as between July and September, although they have been known to ovulate regularly until pregnant. Boars appear to be fertile all year round.

January matings occur shortly after a sow has given birth. Some sows will wander into neighbouring territories in search of a suitable mate in order to ensure the best-quality genes for her offspring or she may decide the best-quality genes are with her home dominant boar. And around this time, the dominant boar will be doing his level best to 'guard' his sows against any wandering lotharios.

A boar will 'churr' to a potential mate and strut around scenting until he entices her from the sett. The sow may be interested in his advances or run off – this depends on her mood! If all goes to plan, copulating can last from a few minutes up to 90 minutes. Some of the shorter matings appear to simply be prompting the female to ovulate.

By the end of April, females will be carrying fertilised eggs – but here's the clever thing... the sows are not yet pregnant and won't be until conditions are ripe. Only then will the fertilised eggs – carried in the womb in suspended development – be implanted, usually December time.

This delayed implantation is clever for a few reasons. Firstly, it ensures the best possible start for the cubs, as they will be emerging from the sett during spring when food resources are picking up. It also means the sow can use the quiet dormant period of winter for gestation, conserving her fat and energy resources which will be converted to milk to nourish the cubs.

Another reason for delayed implantation appears to be ensuring the best possible condition of the sow before she falls pregnant. Should it be a rough dry summer with scarce resources, or illness has set in, the fertilised eggs will be ejected or reabsorbed and the sow will try again the following January. Alternatively, if it's a good year and the sow is in good condition with plenty of fat reserves to see her and her potential cubs through winter, then the eggs will be implanted.

The threats to the potential mother-to-be do not end there, however. Each sow will be determined to give their offspring the very best chance of survival, and to this end, the dominant sow in the clan may hassle and fight other pregnant sows so their eggs fail to implant or they miscarry.

All being well, the pregnant sows will retreat to the setts during late December to snooze away the winter and, after a seven-week gestation period, give birth between mid-February and mid-March (depending on the temperatures).

Litters vary between one to five cubs – but mortality is high. It's not just one or two cubs in a litter that might perish but, often, the entire litter.

Factors include a mother's milk supply being inadequate or failing or the destruction of the litter by a dominant sow.

Cubs are born blind and helpless and their pink bodies are covered in a thin layer of fur. Thanks to mother's nutritious milk, they will grow quickly and within a few weeks their eyes are open and milk teeth have erupted. The cubs will spend a few more weeks clambering around the sett before it's time to take a peek at the big, wide world. Even then, it's only very late at night and with Mum nearby. They'll be around 12 weeks old before emerging with the adults, spending longer periods outside the sett and venturing a little further.

The cubs will spend all of summer and autumn foraging with Mum. They should reach adult size within six months and they will need to reach adult weight by the time winter arrives to have any hope of surviving until the following spring.

Young badgers will reach sexual maturity after their first birthday.

Many badgers will remain with their clan for the rest of their lives, both boars and sows, hoping to eventually gain the dominant position at some point or another. Young boars that attempt to fight their position too often will be expelled and some, both sows and boars, will wander off to join nearby or neighbouring clans. A mix of wandering souls as well as illicit 'affairs' for both sows and boars ensure inbreeding amongst clan members is avoided.

## Keep the Noise Down...!

Forget the silence of the night if badgers are about, as they can be quite noisy! Apart from ambling and bumbling through leaves and dry vegetation, snuffling along the ground, their heads weaving from side to side in search of food, they make for noisy eaters too, slurping up worms and munching noisily on fruits and other finds. Even dry food can be soundly enjoyed!

Badgers also use an extensive range of calls and chatters. WildCRU (Wildlife Conservation Research Unit) carried out some extensive research on vocalisations by badgers at Wytham Wood, a 1000-acre semi-natural woodland owned by the University of Oxford.

Sounds include the 'churr', a deep throated mating call. Not to be confused with the 'purr', a softer noise a sow will use with her cubs particularly during grooming. She'll also use the 'cluck', a soft quack sound with her cubs during play and grooming.

The 'growl', 'snarl' and 'hiss' will be used during more confrontational encounters or in warning. The 'kecker', 'chitter', 'wail' and 'yelp' usually indicate surprise, axiety, distress or pain or sometimes used by cubs during play (particularly if it's getting a little boisterous!).

And then there's the 'grunt' – a low-pitched sound associated with grooming.

A short but delightful recording is available on YouTube with some of the different badger noises the

WildCRU team have collected: www.youtube.com/watch?v=b4lpFjHsGLo.

## Chowtime

Despite being Britain's largest carnivore, the badger's diet includes a wide range of foods, prompting renowned badger expert Ernest Neal to describe the badger as an 'opportunistic omnivore'. By this he meant that badgers will eat whatever is available, whenever it is available – and this includes seasonal variations!

By far and away, a badger's favourite food is the common earthworm (or lobworm) and they will happily snuffle up a couple of hundred a night if available. This moist nutrition will also make up a good portion of their water intake, with badgers only needing to top up from time to time or during dry spells.

Badgers also enjoy slugs and snails, frogs, and insects, such as beetles, bees and wasps. Badgers will happily barge their way into an angry wasp's nest to get at the grubs, barely mindful of the angry stings from the stripey tenants.

Fruits, nuts, acorns and – much to the farmer's annoyance – wheat and corn crops are also considered fine fodder by our lumbering, determined mustelids. Badgers are also known to climb, if necessary, to get at goodies.

Small mammals, such as mice and rats, ground-nesting birds, their eggs and even animal carcasses and carrion, can be devoured during a badger's nightly forage in their territories – areas of which can extend for several miles.

There is some controversy around badgers eating hedgehogs and concerns that this is leading to a decline in hedgehog populations. Badgers are, indeed, one of a hedgehog's main natural predators as only badgers have the strength and dexterity to get past the spines on a hog. However, hedgehogs, despite sharing a taste for similar foods (earthworms and slugs) have been co-existing with badgers for centuries and will actively avoid areas where badgers are likely to roam. As a result, most incidents would only occur as part of a badger's 'opportunistic' feeding nature – if a badger is hungry and there's an available hedgehog...

The more likely cause of hedgehog decline is 'the tidy garden' – leaves and overgrowth swept and tidied away, fencing and walls restricting access and movement, thus lack of available food and shelter as a result, and also the extensive use of pesticides. If, however, you are being visited by badgers, it would be wise not to encourage hedgehogs into the area as well, or vice versa. It makes for an easily available food source for Mr 'What's on the Menu Tonight?' Badger.

'Urban badgers' are becoming more widespread – due, largely, to increased development of areas normally used by badgers. In their fashion, badgers are adapting well and setting up residence in parks, wastelands and cemeteries. Look out for sloping banks and badger paths near where you live – you might be surprised to find a sett!

Badgers habitually forage alone and even actively avoid each other when

snuffling around for food.

They also adapt to the seasons and food availability. In summer, there will be an abundance of earthworms and insects to feed on as well as small mammals. In late summer, they'll take advantage of the ripened cereals on farmlands and in autumn snaffle up dropped fruit, such as apples, and forage the ripened berries in hedgerows.

In autumn, badgers will eat as much as possible in order to build up fat reserves ready for their sleepy winter months.

What to feed Mr Badger if he's visiting you? A good variety will be greatly appreciated – including dog or cat food, peanut butter and fruit. While raw peanuts and mealworms are very much appreciated, too many can cause bone issues. Best to ensure he's not making a nuisance of himself digging up nearby gardens before you encourage their visits to yours!

## Home, Sweet Home

In *The Sword in the Stone* by T.H. White, the eventual King Arthur (nicknamed Wart) is transformed into a badger by Merlin as part of his training and education, as, according to Merlin, badgers are the second wisest animals on the planet (owls being the first). Wart happens upon an old badger, who muses sagely: 'I can only teach you two things – to dig, and to love your home.'

What master diggers and master homemakers badgers are! Of all the mustelids, only Mr Brock will excavate his own home, sweet home rather than taking over other underground burrows or natural

ready-made holes. For prospective badger spotters out there, given badgers will spend approximately three quarters of their lives in their sett, they usually have a standard checklist of requirements for their all-important abode.

The sett will preferentially be started on a slope on edges of woodland or hedgerows, in well-drained, easy to dig soil and the badgers will use their powerful claws to dig out the earth. Entrances themselves will be on bare soil, although there will be plenty of covering around – usually trees and other vegetation. Elder is a particular favourite as an entrance covering and it would seem badgers are very fond of elderberries! Bracken is also a popular covering. The sett location will take into consideration factors such as abundance of nearby food sources, preferably near pastureland or farmland.

The entrances are wider than they are tall – as is fitting of the badger shape – and about the size of a football. 'Spoil heaps' will form part of the landscape around the entrances, containing dug out soil, rocks, old discarded bedding and even bones of past inhabitants. Around the sett entrances will be well-worn pathways used time and time again by the brocks leaving and returning to the sett.

Tunnels descend downwards from the entrances before turning into a series of interlinked chambers and tunnels – varying in sizes and lengths. Nesting chambers are smaller, giving the badgers opportunity to sleep curled up or snuggled together for warmth. Badgers rotate nesting chambers every few days to allow them to freshen up and parasites to die off. Whilst the main latrines will be outside the sett, there will be one or two small ones inside, believed to be used only by nursing sows or during winter confinement.

The sleeping chambers will be filled with plenty of bedding – made up of leaves, straw, dry grass and bracken. In the summer months, some greenery might be brought in as well. Author H. Mortimer Batten wrote: 'The badger is by far and away the cleanest wild animal that we have.' Indeed, badgers are fastidious about airing and renewing their bedding in an effort to keep parasites down and have been seen dragging new bedding to the sett from a considerable distance, disposing of the old.

Mr Badger will be sure to clear his sett of any decaying foodstuffs or other unpleasantries.

In some wonderful discoveries, it's been found that badgers are cleverly able to use local materials to ensure a better night's sleep – including the use of plastic bags and fertilizer sacks as a makeshift damp course to line nests before covering them with traditional nesting material.

One study of badgers kept in an artificial concrete sett recorded them chopping up a mix of hay and straw into a pile and allowing it to ferment. The core temperature of the pile rose to 38°C. The captive badgers moved their nest closer or further away from this natural radiator as they needed. In addition, badgers have been known to block sett entrances in colder temperatures. It all makes for a very cozy badger home!

The main sett will be permanently occupied by between 2 and 20 badgers. Annexes or 'overflow' setts will be dotted around just a short distance from the main sett and are used regularly, if the main sett is busy with young cubs, for example. Further afield will be subsidiary setts, which work for overflow like the annex setts but are a little further away, and on the outer reaches of a badger's territory will be the outlier setts – usually with just the one entrance and used only occasionally. Most badger clans in England have between 3 and 6 setts in their territory. The size of the clan and the territory it maintains is dependent on available food; the more abundant the food sources, the bigger the clan and territory.

Studies have found that badgers move happily from sett to sett but will rarely stray into a neighbouring territory. Badgers defend their territories fiercely and, if an interloper is happened upon, fighting is likely – particularly during the peak mating season.

The sett is a very important part of a badger's world – they will normally head home at daybreak and head out for grooming and feeding soon after dusk. Whilst badgers do not hibernate during the winter months, they do prefer to remain underground for longer periods, using up their body fat reserves built up during autumn and slowing down their metabolism and heart rate and even their temperature and breathing – a condition known as torpor.

Setts will be occupied, expanded, mended and maintained by successive generations of badgers and they will continue to occupy a sett, if undisturbed, for hundreds of years. And it's not unusual to find other animals, such as foxes, rabbits and mice, that have moved into disused entrances and tunnels.

The Protection of Badgers Act 1992 expressly forbids interfering or obstructing a badger sett. As such, when badgers have elected to excavate their home somewhere entirely inconvenient for man, such as under buildings or roadways, the setts must be carefully closed down and relocated by professional badger consultants and only with official permission.

## Off to the Loo

In keeping with H. Mortimer Batten's assertion that badgers displayed an 'idea of healthful sanitation', badgers are well known for their latrines – their 'outhouses'. Although, as previously mentioned, small latrines will be found in the sett during cub season and cold weather, these shallow holes (between 5cm and 20cm deep) are usually found spaced around the different setts as well as further afield along their territory boundaries and, largely, around favourite feeding areas.

In colder weather and during breeding season, latrines will often be found closer to the sett entrances.

Badgers leave their poo uncovered in the latrines, as they serve as scent markers for the home badgers as well as deterrents for badgers straying into unwelcome territories.

# Badger Watching

Probably the easiest way of spotting badgers in their natural environment is booking yourself onto one of the many badger-watching experiences available around the UK. This will enable you to sit in the relative comfort of a hide at a location where badgers are almost always (never 100%!) guaranteed to appear. In some cases, floodlights are used as well as encouragement in the form of peanuts and other goodies scattered nearby. The badgers have become accustomed to both and will venture relatively close to the hide, offering some wonderful viewing opportunities.

If you're out on an amble in the countryside, you might be lucky enough to spot a sett. Look out for the clues. Firstly, bare patches of earth on a slope with sizeable holes will be a good start. The holes should be roughly half-moon shaped. Round, smaller holes will usually indicate rabbits whereas tall-ish, narrow holes will indicate foxes, especially if bones and food detritus are visible near the entrance. A tell-tale whiff will also indicate a fox hole rather than a badger sett.

Nearby the badger-shaped holes you should find a steep spoil heap of dug out soil, stones, rocks and discarded bedding. There might also be some scratching posts evident nearby – trees, often elder, with bare or well-scratched patches on the trunk up to around a meter high from the ground.

Then, keep an eye out for the nearby latrines – conical shaped depressions in the ground filled with uncovered poo. Add some clearly defined pathways around the area

and you can be fairly sure you've found Mr Brock's home! If you're extra lucky, you might even find some paw prints nearby, depending on the ground surface. The paw print will be wider than it is long and will show the depressions of several of the toes – four or even all five – as well as narrow claw marks.

Badger watching in the wild requires a degree of planning and preparation. First, visit the sett during the day to get a feel of the lie of the land and perhaps work out the best place to sit for your watching experience. This spot should be at least 10 metres away, if not a little more. Badger eyesight is quick to spot shapes that don't belong and their hearing is sound enough to tolerate any regular nearby noises but will quickly pick out noises that are unusual. Try to wear dark, rustle-free clothing and get as comfortable as possible in your chosen seating position. Fidgeting, coughing and sneezing will scare the badgers right back into their setts, where they'll wait, quite patiently, for you to clear off!

Wind and weather will be your friend or foe. ALWAYS sit downwind. A badger's nose will pick you out faster than you know if the wind carries your scent to them. Experienced badger watchers will tell you that

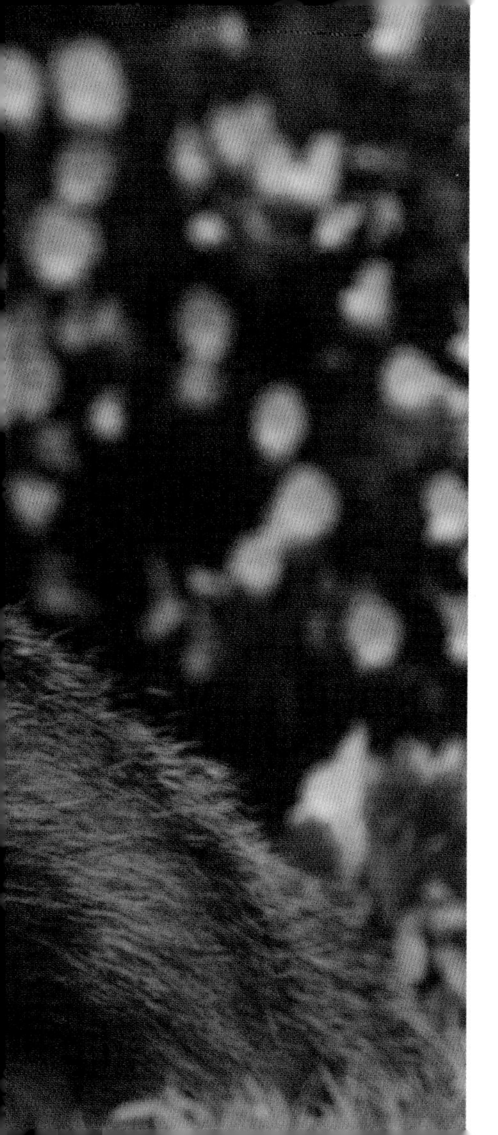

cold easterly winds usually mean badgers will stay tucked up in bed for most of the night. A warm, damp night, however, and badgers will be out as soon as possible, snuffling around for their favourite lobworms.

Should you be lucky enough to have urban badgers nearby (and you're not trying to encourage hedgehogs as well!), leaving out some healthy snacks, such as dog food and raw, unsalted peanuts, would be a nice top-up to their normal nightly foraging. Please remember, however, badgers are wild animals and should be left to be wild.

# Threats to Badgers

## Far and away the badger's biggest natural threat and predator is man.

In Britain it is thought that only red foxes, buzzards or golden eagles would have the size and strength to bring down a badger – but, even then, the badger would have to be sick or wounded or very young.

Bring man and their cars into the equation and you've got a different situation entirely.

The Badger Trust reports that around 50,000 badgers are killed each year on the roads – that's believed to be around 20% of the badger population. This killing spree seems to peak in spring so it's thought to be mostly adult badgers seeking out mating opportunities.

Under licence, for a six-week period each year, badgers are culled in a selection of bTB 'hotspots' with the aim of targeting 70% of the population in that area. Since 2015, over 125,000 badgers have been killed. Whilst the International Union for the Conservation of Nature (IUCN) currently lists badgers as an animal of 'least concern', one wonders how much longer it'll be before badger populations begin to suffer, particularly in light of hotter summers, meaning a bad time for badgers and badger cubs due to dehydration and starvation.

Illegally, badgers are also falling foul of snares, poisons and even domestic dogs to control 'vermin', forgetting the role badgers (and foxes) play in keeping down rat and other small mammal populations.

Then there's the destruction of habitat – rapid loss of viable land for new setts as well as destruction of old setts due to industrialised farming and building development.

Thankfully, urban badgers are coping well, largely due to plentiful food supplies but their setts can wreak havoc if they're dug under roads or along railway lines or any other areas that could cause damage to buildings or properties. Under licence, these setts will be blocked and, if possible, moved.

If man isn't enough to contend with, there are the 'natural' threats to badgers, of which their biggest are badgers themselves! Fiercely territorial, injuries sustained through fighting can easily be fatal, if not the fight itself. Wounds that fester can leave the badger crippled and unable to feed or defend itself. In addition, it's not uncommon for a dominant sow to kill the cubs of subordinate sows in order to ensure the best chances for her own cubs. Badger-land is tough out there!

Other natural threats to badgers include parasites – those ingested while snuffling around in the dirt as well as those hitching a free ride, like ticks, fleas and lice. Impeccable grooming in healthy badgers helps keep the external parasites under control. Each evening, a badger's first task upon ambling out of their sett is a routine but thorough clean-up of themselves and others in the clan as well as a regular clear out and replacement of bedding from the nests.

Fortunately, diseases such as rabies have been wiped out in the UK and cases are very rare in Western Europe. However, bTB passed on from cattle can infect badgers, although the numbers of badgers diseased or killed by the virus are sketchy. Some badgers manage to avoid the infection altogether whilst some are infected but not contagious. Some are contagious but show no symptoms and then you get the unlucky ones that succumb to the horrible disease, show all the symptoms and are highly contagious. Unfortunately, despite the large numbers killed during the cull, very few are tested post-mortem, so the definitive numbers of the disease in badgers is still unclear.

## And Then There's Badger-Baiting...

Synonyms for the verb 'to badger' include pester, harass, plague and torment. The word usage is believed to originate from the late 18th century from the 'sport' of badger-baiting, a vile and cruel pastime in which badgers are dug out of their setts and pitted against dogs for a fight to the death. Indeed, dogs have been specially bred – terriers and also dachshunds (literal translation from German: 'badger hound') – to be adept at running into the badger sett tunnels and attacking the badger within its own home, before both are dragged out to continue the fight in a confined space. This despicable brutalisation of both badger and dog has been going on for centuries and continues to this day despite the legal protections in place for both animals.

In his wonderfully detailed and thoroughly researched book *Badgerlands*, Patrick Barkham puts forward the reasoning that whilst kings and gentry hunted foxes and deer on horseback, the common man was left with the badger to hunt for their 'fun'. The badger's fearsome tenacity for defending itself

is considered 'good sport' for the betting man to watch pitted against another innocent animal.

Even the Cruelty to Wild Animals Act in 1835 was not protection enough at the time, for it gleefully used the loophole of the wording 'wild'. It was still legal under the act to dig a badger out as long as dogs were not set upon it, so badgers were instead bagged and held captive as 'tame', only to be trotted out during games and used for as long as it could withstand the attacks.

Badgers also fall foul of the dreadful practice of 'lamping', in which, during the night, powerful torches or lamps are suddenly turned on with the intention of dazzling the prey – such as rabbits and foxes – before shooting them or releasing dogs like lurchers to chase after and catch them while they're still momentarily confused.

The Protection of Badgers Act of 1992 put in as stringent controls as it could, but badger-baiting and lamping continues to this day – albeit illegally. Persecution of badgers is currently one of six UK Wildlife Crime priorities.

Should you see or suspect badgers being dug out or their sett being interfered with, first ensure your own safety and then please call the police immediately – it is a crime.

In addition, please log all findings – new setts and badger traffic casualties – as well as any suspicious activity with The Badger Trust at www.badgertrust.org.uk/report. If possible, report badger traffic causalities to your local badger groups, as they will have

volunteers available to check the animal is indeed dead rather than suffering, as well as locating any nearby setts to potentially rescue young cubs.

# The Cull

The badger cull is a wholly contentious and emotive issue on both sides of the divide, with passions running high for those opposed to the cull as well as for those affected, living and working in the dairy industry.

It is also, as some believe, more about politics, with badgers caught in the middle of political parties wrangling for votes.

Let us start at the beginning. What's this cull all about and what is bTB?

bTB is a disease caused by a bacterium (*Mycobacterium bovis*) that's closely related to the species responsible for human tuberculosis (*M. tuberculosis*). It is a bacterial infection which enters the body through the nose or mouth to join the bloodstream and travel to organs like the lungs, kidneys, liver and intestines. The disease forms lesions in the organs themselves and, as such, is very difficult to spot in the early stages.

Thanks to the BCG vaccine and antibiotics, countless human lives have been saved from this terrible disease but it was a scourge for centuries, peaking in the 1800s when 'consumption' caused an estimated 25% of all deaths worldwide. Drug-resistant strains began to appear in the 1980s, however, and new infections occur in about 1% of the world's population each year.

Tuberculosis has been found in cattle in slaughterhouses since the early 1800s. In 1971, a post-mortem was performed on a dead badger found in Thornbury, Gloucestershire, an area suffering a high bTB rate. The badger was found to be carrying bTB and this was the first known evidence

that the disease had not only jumped from humans to cows, but had now jumped from cows to badgers.

It is thought that urine and faeces are the most likely culprits in the transmission of the disease. The disease could also linger in the soil (under infected cowpats where badgers might find beetles and juicy worms) and even in the water.

More post-mortems were carried out on more dead badgers and, while most badgers were found to be disease-free, a further three were carrying the infection.

This resonated with farmers who believed that the increase in the badger population coincided with the rise in bTB cases.

Badger populations were thought to be on the up since the Badger Act came into being in 1973 and the subsequent 1992 law criminalising any interference with setts. Badgers were finally protected and wildlife crime officers were doing their best to prevent digging out and baiting of badgers and prosecuting the perpetrators where they could.

Climate change also appeared to be beneficial for the badgers. Warmer winters meant badgers were living longer and breeding longer and more litters were surviving the lean times.

Modern agriculture offered plentiful crops for the animals, including maize, which would ripen in autumn. Perfect timing for badgers to fatten themselves up to survive the winter.

In the 1930s, around 40% of cattle across Europe were found to be infected with bTB. Thanks to the pasteurisation of milk, human deaths from the disease were reduced to almost nil but the effort was on halting the cow-to-cow infection.

During the 1950s, compulsory testing was introduced and infected animals were slaughtered. Movement restrictions were in place for untested or infected herds and during this time the disease steadily declined.

And all might have been OK for our friend the brock, but for that infected badger found in Thornbury. This set up a flurry of excitement and the government issued licences to permit the gassing of setts, since

the Badger Act itself stated that badgers could be killed to 'prevent disease'. This gassing of badgers, met with a wave of public opposition and a 'Save our Badgers' campaign was run ahead of the 1979 elections. The newly elected Conservative government halted the gassing and then banned it permanently in 1982, when more evidence was coming to light that the gassing method did not always deliver a 'clean kill' and that badgers were being left to die slow, agonising deaths.

It went relatively quiet during the 80s and 90s. Badgers could be killed under licence if found on infected farms but, at that time, the bTB numbers were dropping as well, down to just 0.5% in 1981.

Unfortunately, the foot-and-mouth epidemic of 2001 started the increase and spread of bTB cases in cattle all over again. Entire herds were wiped out in an effort to control the spread and, once contained, farms were desperate to restock. Cattle were moved across farms and across counties with only minimal checks to restock farms as quickly as possible. The number of bTB infections skyrocketed from a little over 5,000 cases in 2001 to over 20,000 by 2004.

Others believed that the intensification of dairy farming as well as the selective breeding to produce higher milk yields meant the cows were simply more susceptible to the disease. Animal husbandry was producing weaker animals more likely to become infected.

In spite of the obvious links with the rapid restocking of herds and the questions around weakened breeds,

talks about killing the badgers to halt the spread of the disease started up again.

In 2007, the Independent Study Group on Cattle TB concluded that badger culling would make no meaningful contribution to the control of bTB in cattle and some of the policies being considered could actually make matters worse. For example, destabilising groups/setts and individual badgers moving into other areas, increasing the chances of an infected badger coming into contact with cattle. This is known as the 'perturbation effect'. Their report also concluded that the disease could not only be contained but also REVERSED using rigid cattle-based controls. Badgers could be left entirely alone.

The Labour Government, at the time, heeded the advice and badger vaccination trials were started in South West England. Unfortunately, the election of 2010 saw the formation of a Coalition government and the decision reversed. Mr Brock was suddenly back on the hit-list.

The culls started in 2013 and continue each year for a (roughly) 6-week period, with new areas added annually. In 2016, 7 new areas were included. A further 11 in 2017. Yet another 11 in 2018 and 40 new areas in 2019.

A total of 19,274 badgers were culled in 2017. Over 32,000 badgers were killed in 2018. Targets for 2019 are a staggering 64,000 badgers.

Methods are free-shooting or trapping and then shooting. Reports on the efficacy of the culls are varied, depending on which newspaper

or publication you read. It's totally baffling. Pick your statistic to prove your point of view.

What remains an issue is that whilst there is in fact a vaccine for cattle, there is no designated certification for animals of 'vaccinated but free of bTB'. A positive result of a later test on an animal may actually be a response to immunisation.

Without the political will to rectify this, it could take a decade to resolve.

The current test for bTB is alarmingly inaccurate, producing 'false positive' results in which reactive cows believed to have the disease and therefore slaughtered are regularly found to be disease-free when subsequently tested by a lab. 'False negatives' are just as worrying, an infected herd being given the all-clear.

Farms that are shut down suffer economically. Compensation offered for each slaughtered cow rarely covers the true cost of the animal and doesn't take into account the financial disruption caused to a farm by having to keep the animals longer than planned during their restricted movement period. If a herd is found to be infected, they must be tested every 60 days until a minimum of 2 clear rounds declares them disease free. Or, they are slaughtered.

Badger vaccination programmes are in place in some parts of England and Wales with a view to eliminating the disease in the badger. There is, as yet, no conclusive evidence that the badger IS transmitting the disease back to cows... only that they CAN transmit the disease. One recent study showed that badgers would prefer, on the whole, not to get too close to cows.

bTB has been found in other mammals too, including deer, foxes, mice and even cats and dogs, but the most likely spread of the disease is from one cow to another. Movement of infected herds and even the spread of infected slurry would be simple contributory factors.

Opponents of the cull believe the badgers are but a political scapegoat, that the government is keen to prove to farmers that it is 'doing something' about the disease in a thinly veiled attempt to garner votes from the farming and countryside communities.

Finally, there is some question as to the 'size' of the bTB problem versus the incredible number of badgers being killed. A survey in 2011 revealed that of 20,000 dairy cows prematurely killed, the largest percentage was due to the cow not being in calf, with mastitis and other related infections causing poor quality milk. Lameness, age, accidents and others (including 'died in a field') made up the rest, with just 3.23% being related to diseases such as bTB and bovine viral diarrhoea.

At the time of writing, 331,972 cattle have been slaughtered in England as a result of bTB since 1st January 2008 – that's over 12 years.

Once the cull targets are met by the end of this year's killing season, over 128,000 badgers will have been slaughtered in just 6 years.

It should be noted, Scotland has badgers but no bTB. Furthermore, bTB has sadly been identified on the Isle of Man. There are no badgers on the Isle of Man.

In March 2020, the government announced its intention to phase out culling 'in the next few years' in favour of vaccinating both badgers and cattle (once a reliable vaccine has been found for cattle). No dates have been put on the announcement and animal rights groups fear the announcement was lip-service in light of the continued fierce resistance to the culls.

Time will tell. Let us hope it won't be too late for Mr Brock...

cheshire
wounded
badger
patrol

# The Badger Protectors – Wounded Badger Patrol Cheshire

In 2017 the badger cull spread to Cheshire and a very determined, experienced and organised group met it head on.

Neil Copping, a seasoned patroller from culls in other parts of England, quickly formed the Wounded Badger Patrol Cheshire and, alongside Jane Smith, deputy leader of the Animal Welfare Party, threw all the resources they could at it, including a well-manned and informative Facebook presence that now has over 1,700 followers.

Every night of the duration of the cull licences (roughly six-weeks), groups of volunteers head out to patrol their extensively researched routes. That's what Wounded Badger Patrols do... they travel along public right of way paths and footpaths and look out for any signs of wounded badgers,

The author protesting the cull outside Westminster, 2015.

inexpertly shot but not killed and needing medical attention.

The groups are law-abiding and non-confrontational, giving everyday

people like us an opportunity to actively oppose the badger cull. Their presence alone is often enough to drive the shooters away from a particular area. Indeed, the shooters are not allowed to free-shoot whilst the public are walking nearby.

The free-shooting method has come under a lot of criticism from welfare groups, as badgers risk being shot but not killed outright, leaving them to suffer and die slowly.

The shooters are allowed, however, to cage-trap the badgers, with strict rules in place that the badger must be shot before midday so as not to leave the animal suffering, trapped in a cage without food or water. Patrollers keep an eye out for the trapped badgers too and any badgers left too long in the cage or any setts found to be illegally blocked or tampered with are reported to the police.

The patrollers meet up just past sunset and walk throughout the night along regular and pre-arranged routes. There are even dedicated patrollers out checking their routes before heading to work!

Behind the scenes, volunteers spend hours organising patrol leads, meet-up times, lift-shares, online updates and visible protests. It's a very busy team!

In my experience of walking with the Wounded Badger Patrol, I've been amazed at the warmth, passion and camaraderie of the patrollers, people from all walks of life, travelling in from near and far. Neil, for example, lives in North Wales, a goodly drive to and from the meeting points in Cheshire each evening.

In the words of Jane Smith this is a '100% peaceful, 100% lawful vigilant presence offering civil resistance to this shameful cull. We have more people joining us every year and we always welcome new

volunteers. Our volunteer age range is currently 18 to 83 and our ranks include grandparents, home-makers, students, GPs, teachers, musicians, ecologists, vets and lecturers.'

The Cheshire cull zone is made up of four areas, two of which are totally within the county and called Cheshire and two other zones which encompass Shropshire and Staffordshire. DEFRA's total target for the area in 2019 was 3,500 dead badgers.

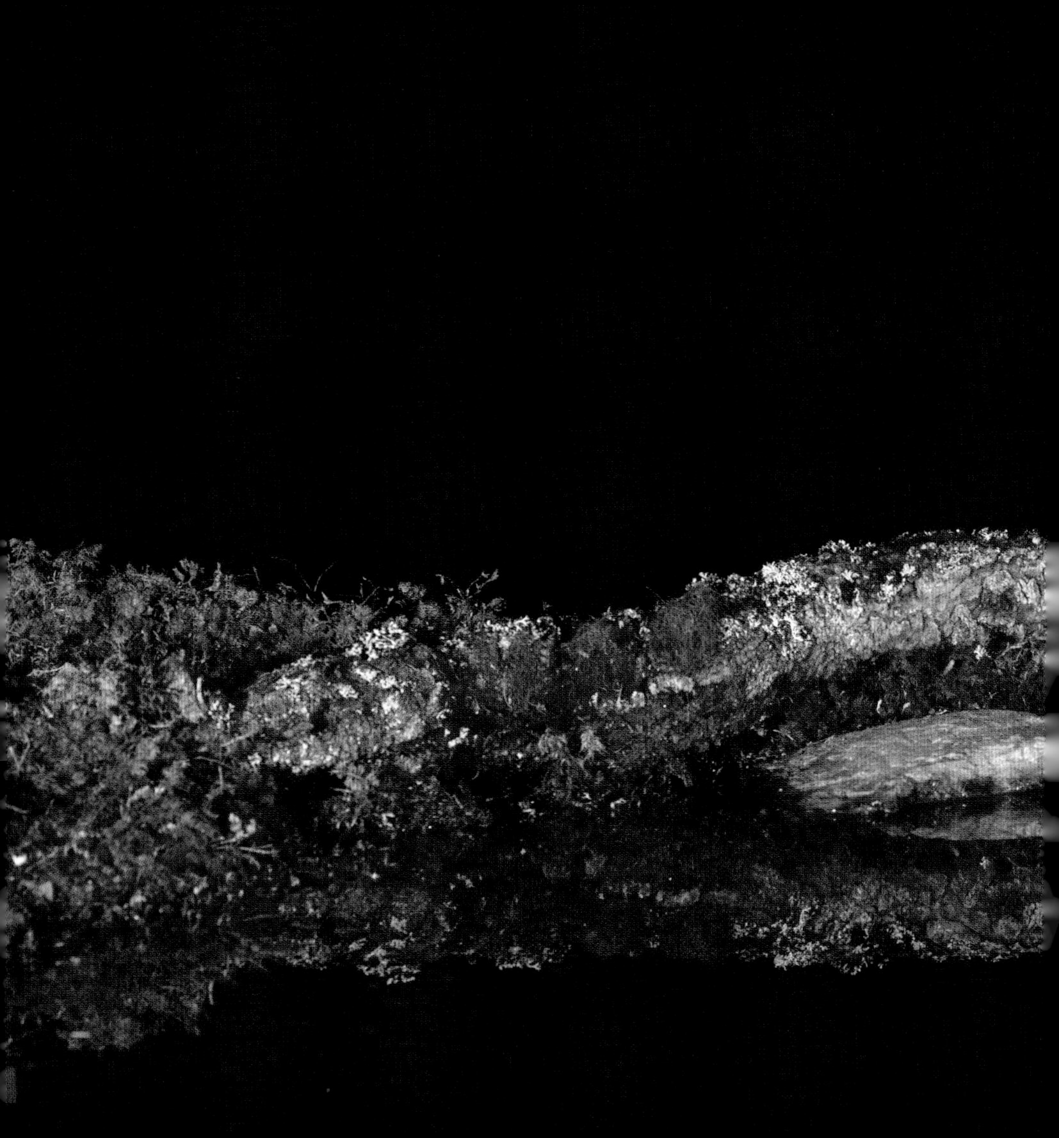

The team are under no illusion that the shooters will meet their targets each year but they have ensured that any wounded badgers will not have to suffer a moment longer than necessary. At next cull season, they'll be back, prepped, checked, mapped and trained again. They'll be ready and waiting.

You too could join a Wounded Badger Patrol! If not the team in Cheshire, then a patrol group working in a cull zone near you. If you're over 18, all you'll need for your kit is: a good pair of walking boots, a hi-vis, a powerful torch with plenty of spare batteries and a fully charged mobile phone, to stay in touch with the groups. Bring along a snack if you like and some waterproofs during dodgy weather! For the more experienced walkers/hikers, your night vision goggles and body cams would be much appreciated. With a dollop of enthusiasm and determination, you could help save the badgers!

In addition, there's an opportunity to meet some wonderful like-minded people and enjoy an evening out in the countryside, perhaps spotting an elusive owl or other wild animals and saying 'hello' to a cow or two along the way (always curious, they are!).

To get in touch with the Wounded Badger Patrol Cheshire team, email wbpcheshire@mail.com or you can find them on Facebook.

For the most comprehensive list of Wounded Badger Patrol groups across England, visit the Badger Action Network: badgeractionnetwork.org.uk.

wound
badger
patrol

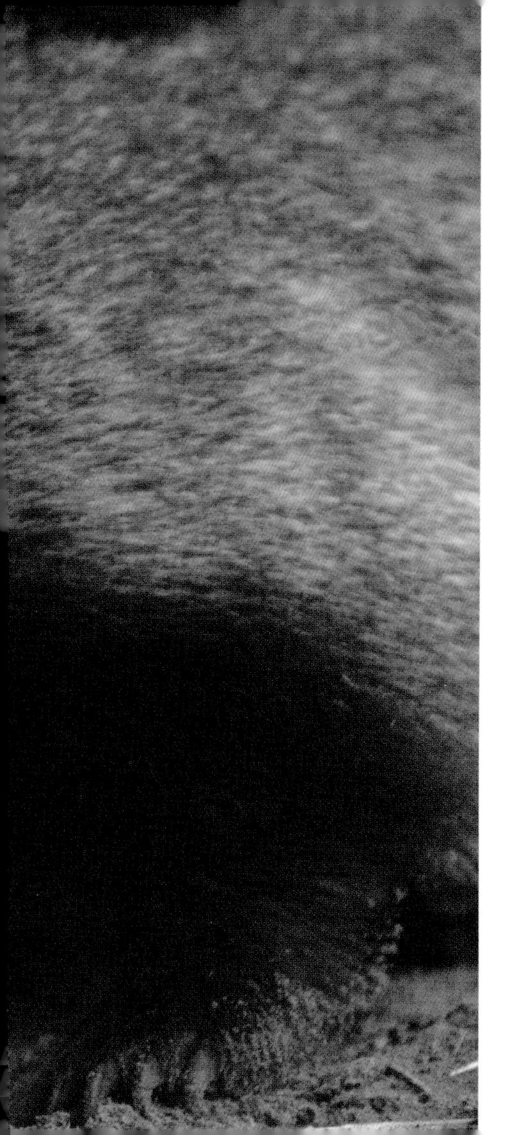

A shapeshifting badger from the *Wakan Sansai Zue* (Illustrated Sino-Japanese Encyclopedia), 1712.

# Badgers in Myth and Legend

From the world of the superstitious, an old remedy to ward off the ghosts and ghoulies suggested the following:

*A tuft of hair gotten from the head of a full grown brock*

*is powerful enough to ward off all manner of witchcraft;*

*these must be worn in a little bag made of cat's skin – a black cat*

*and tied about the neck when the moon be not more than seven days old*

*and under that aspect when the planet Jupiter be mid-heaven at midnight.*

An amulet that wasn't for the faint-hearted – and certainly wasn't much luck for the brock or the cat! It is a 'remedy' that does appear, however, to stick in people's imagination, even quoted by Lord Gisborough during the reading of the Badgers Bill in February 1973.

It was reported as late as 1800 in the *Sporting Magazine* that the flesh, blood and grease of a badger were useful in oils and ointments to cure ailments such as coughs, colic, kidney stones, shortness of breath and even sprains. It was said that badger grease was so penetrating that, when rubbed into hands, the grease would come right out of the back of the them!

Your luck on encountering a badger was either wonderfully good or woefully bad depending on whether they crossed in front of you or behind you:

*Should a badger cross the path*

*Which thou hast taken, then*

*Good luck is thine, so it be said*

*Beyond the luck of men.*

*But if it cross in front of thee,*

*Beyond where thou shalt tread,*

*And if by chance doth turn the mould,*

*Thou art numbered with the dead.*

*And your days were numbered too if you were to hear a badger call followed by an owl according to the old rhyme:*

*Should one hear a badger call,*

*And then an ullot cry,*

*Make thy peace with God, good soul,*

*For thou shall shortly die.*

Have you ever heard the expression 'rough as a badger's bum', either referring to the effects of the night before, an unkempt or unshaven person or rowdy behaviour? This seems to refer to the badger's propensity for biting each other's rumps during fights. Most badgers seem to wear scars around their tail and rump area – including females – where chunks of fur and skin can be torn off in scuffles.

Do badgers really bury their dead? Myths about badgers walling up their dead in the sett or dragging bodies back to the sett appear to be just that – myths. As we've already ascertained, badgers are not particularly sociable, despite living together, so would have little or no need or real interest in burying their dead or retrieving dead or dying clan members. Stories of badger bones found in walled up parts of their sett are, unfortunately, probably due to the period of time when gassing of badger setts was legal. Once the gassing process had started, any leakages seen coming out of the sett were blocked to ensure the gas completely infiltrated the whole sett. Fortunately, gassing is now illegal.

In Japan, depending on the region, the Japanese badger (*Meles anakuma*) is called a 'mujima' or 'mami'. Japanese folklore often depicts badgers as 'yokai' (ghosts) or shapeshifters in the form of mischievous women or even monks. In several legends there is often confusion over whether the tale is about a badger or a 'tanuki' (racoon dog) and subsequent telling of the legends over the centuries blurs the distinction.

## The Teapot Badger (Bunbuku-Chagama)

There once was an old man, poor but kind, who liked to help the local children, sharing what food he could and teaching them to draw and write. Badger was passing one day and saw the old man and thought, 'I shall repay his kindness', and he turned himself into a teapot full of tea. The old man was utterly delighted when he found the teapot, as he'd never been able to afford one of his own.

That evening, when the children arrived at the old man's home for a story, he showed them his new teapot before heading to the orchard to pick fruit for his guests.

The mischievous children decided they would steal the old man's teapot. However, as they tried, the teapot suddenly turned back into Badger, who wasn't best pleased, and, with flashing sharp teeth and a tail waving in fury, he nipped at the children, who ran from the old man's house, shrieking.

Badger teakettle, Japan, mid-19th century.

'The Kettle of Good Fortune' from *Buddha's Crystal and Other Fairy Stories* by Yei Ozaki, published 1908.

The old man rushed up to them and asked what the matter was. The children shouted that the teapot was really Badger and had tried to attack them! By the time the old man reached his house, Badger had turned back into a teapot. The old man admonished the children, saying they were just dreaming, and went back to the orchard.

The scallywags tried stealing the teapot yet again... and, yet again, the teapot turned into Badger, who chased after them, nipping as he went. This time, the old man arrived in time and thinking Badger was attacking the children, tried to hit him. Badger told the old man what was happening and that the children were trying to steal the teapot. The children hung their heads in shame.

By this point, Badger had had enough of cheeky children and suggested to the old man that they travel far and wide, putting on magic shows where Badger would turn into a teapot and back again. The old man agreed and he and Badger are very successful in their endeavour. The old man returns home very wealthy. On the top of the hill in his orchard, the old man creates a little temple for the teapot, where it is greatly revered.

A variation of the story appeared as *The Wonderful Tea Kettle* by Mrs T. H. James in 1866, although her version does suggest a racoon dog rather than a badger.

## The Badger's Money

Another wonderful old Japanese tale tells of a poor monk who spent his days in prayer and meditation. His neighbours respected and admired him so ensured he had enough food to eat and would repair any issues with his modest home.

One very cold night, he heard calling from outside and was surprised to see a badger standing in the snow. The badger respectfully bent its knees and asked the old monk if he might be allowed to spend a short while near his warm fire. The monk ushered the badger inside his home. As promised, the badger stayed but a couple of hours warming himself – then thanked the monk with deep bows and went on his way.

The following night, the badger returned with the same request and the monk invited him in. And all through the cold winter, the badger returned each evening. Soon the monk began to look forward to the badger visiting, for the monk was all alone.

As soon as spring arrived, the badger thanked the monk for his kindness and disappeared. The monk missed the badger so it was with great joy that, as soon as winter's cold breath arrived, so too did the badger. This went on for many years and the badger and the monk became firm friends.

One day, the badger asked the monk what he could give to the monk to repay his generosity. The monk explained that he wanted for nothing, as the villagers were good and kind to him and ensured he was looked after. The badger pressed him for an answer

Mrs T.H. James (Kate James), author/translator, 1845-1928; Yoshimune Arai 1863-1941. Illustration from *The Wonderful Tea-Kettle*, Japanese Fairy Tale series, 1886.

and eventually the monk admitted that he had often longed for three gold coins he could offer at the holy shrine so that prayers and masses would be said for him and he would enter into salvation when he died.

The badger thought for a while and then bid his monk friend farewell and trotted off into the night. The following night, the badger did not return. Nor the next night. Nor for three long years. The monk was very sad and missed his friend.

Suddenly, one night the monk heard the badger calling from outside and rushed to open the door. His furry friend trotted indoors, warmed himself by the fire and presented the monk with three gold coins.

The badger explained that he could have easily stolen any amount of gold for his friend but worried that ill-gotten gold offered to the temple would mean that his friend would never be allowed into heaven. So the badger had travelled to the island of Sado and spent three long years sifting through the sands discarded by other prospectors until he had enough to make the three gold coins in order to thank the monk for his kindness.

For as long as the monk lived, the badger and the monk would spend the rest of their evenings together.

## Native American Mythology

In Native American mythology, the badger is seen to have healing, protective and even restorative powers, particularly amongst the Hopi, Zuni and Pueblo tribes. Indeed, Pueblo tribes believe the badger helps the people and animals reach this world from another one by digging through the rock that seperated them. Arapaho tribes ascribe healing and use of medicinal plants to the badger and badgers are seen as protective parents to their cubs, hard working and cautious, as portrayed in an old Lakota legend:

## The Badger and the Bear

At the edge of a forest lived a large badger family. Father Badger was a great hunter and every day would return with food. Whilst the cubs played, Mother Badger would hang thin slices of meat to dry before packing them away in brightly coloured bags.

One day, as Father Badger sat making arrows while the cubs played nearby, there came heavy footsteps outside the door. Suddenly, the door was thrust open and a dirty, shaggy bear shoved his way into the house then sat quietly in the corner. The bear stared hungrily at the bags of dried meat.

The kindly Father Badger realised Bear was starving and quickly arranged a great meal to share. Bear ate his fill then smacked his lips in appreciation and left.

The next day, Bear returned. And the next day, and the next. Each time he would barge into the Badger home and sit in his corner. Before long, Mother Badger put a rug out for Bear to sit on for his meals.

Soon, shaggy Bear began to look healthy and shiny and strong again. Then one day, Bear entered the Badger home with a wicked glint in his eye. He waved his fearsome paws at the Badger family and roared, 'I am STRONG! Be gone!!' and, with that, proceeded to hurl Father Badger, Mother Badger and the cubs out of their home and into the dust outside.

Father and Mother Badger made a small shelter as best they could to protect the cubs and themselves. Father Badger went hunting but without his arrows he returned to their meagre dwelling empty-handed. The cubs cried with hunger.

Father Badger hung his head and said to Mother Badger that he would beg for food and went to Bear. When he arrived at what had been his house, now filled with Bear and Bear's family, his pleas for food were laughed at. 'Be gone!' roared Bear and booted Father Badger back outside.

The next day, Father Badger tried again. Again, Bear roared and kicked the hungry badger out of the house into the dirt. This time, however, as Badger lay in the dust, he spied a blood clot from the remains of a buffalo Bear had killed and dragged home. As quickly and as quietly as he could, Father Badger gathered up the blood and hurried back to his family in their hovel.

Father Badger resolved to ask The Great Spirit to bless it and he built a pile of sacred stones and heated them. When they were ready, he placed the blood nearby and sat next to the stones and asked The Great Spirit to bless the small amount of buffalo blood.

Father Badger felt a presence at his shoulder. He turned and was astounded and delighted to see a Lakota brave dressed in buckskin. The warrior carried an arrow and had a quiver slung across his back. Father Badger welcomed him joyously and Brave greeted him as his father and announced he was Father Badger's avenger.

Father Badger was proud to have the Lakota warrior call him father but he despaired that he had no food to offer or even feed his own family. 'I will go beg again,' Father Badger announced.

'I will go with you,' said The Brave.

Bear saw the two coming towards the house and was startled to see The Warrior carrying the arrow. He knew this was the avenger of which had been foretold many moons ago.

Indian brave.

He rushed to greet the pair. 'Hello Badger, my friend! Here, take my knife and cut yourself the best bits from this deer I have hunted.'

Father Badger was surprised but he took the knife from Bear. The Brave stepped forward and looked Bear in the eye. 'I am here for justice. You have returned but a knife to my father. Now return his home.' Bear shook in terror of the avenger with the arrow that had been prophesised. He bellowed in fear to his family to leave the home immediately and they fled to the forest.

The Badger family returned to their home singing and rejoicing. Father Badger thanked the Lakota Brave. The avenger left them and went on to travel the earth.

## Celtic Mythology

Celtic mythology also purported badgers to have shapeshifting qualities, turning from human-form into badgers in order to explore the forest. The Old Irish word for badger is *tadhg* or *tadc* and thought to have Gaulish origins. The foster father of Cormac, the great king of Munster in Gaelic Ireland during the 2nd century AD, was named Tadhg and he refused to eat some badgers which Cormac had hunted and brought home for dinner, because Tadhg believed them to be his transformed cousin and kin.

Badgers also appear to be synonymous with virility in some parts, with the badger penis bone – or baculum – being worn as tie-pins or gifted to young grooms.

In the *Medicina de Quadrupedibus*, an old Middle English translation of a compendium of cures and ointments for a range of ailments, a list of uses for a badger – from teeth to testicles – is documented, including: 'If anything of evil has been done to any one so that he may not enjoy his sexual lusts, let him boil a badger's testicles in running spring-water and in honey and let him take it then fasting for three days; he will soon be better.' The text goes on to suggest using the teeth to ward against any kind of evil, boiled badger brain as a remedy for headaches, burying the liver at land boundaries for defence and using badger hide in shoes to ensure comfy feet! Even to this day, badger hair shaving brushes are considered the very best.

## Saint Piran and the Badger (Circa 5th Century)

St Piran, long before he became the patron saint of Cornwall, was said to have been thrown off a cliff edge with a millstone around his neck under the orders of the King of Ireland, who feared his mystic powers. The tale goes that the seas immediately calmed and St Piran floated upon the millstone to the shores of Cornwall.

Arms of the Saint-Peran family (Brittany).

There, at Perrenporth, he built his chapel and his first disciples were said to be a bear, a fox and a badger.

Other legends have it that Saint Piran was also Saint Ciaran of Saignir, born in the fifth century in West Cork. He was renowed for his gift of communicating with animals and a wolf, a badger and a fox would help him and his monks build their huts. One day, the fox stole Saint Ciaran's shoes. The monk sent faithful Badger off to find Fox and convince him to return. Badger did just that and Fox returned repentant.

Whilst the Cornish flag is said to have derived from smelted tin rising to the surface of black stone, one can't help but notice the similarity to the markings of our dear friend brock!

A shapeshifting badger from *Konjaku Gazu Zoku Hyakki* (The Illustrated One Hundred Demons from the Present and the Past) by artist Toriyama Sekien, published circa 1779.

Leather journal by Skyravenwolf.

# Badgers in Art and Literature

Detail of back of badger journal by Skyravenwolf.

In literature, authors often depict badgers as private, sometimes mysterious, ferocious defenders of their home. Loyal and dependable.

## Badger by John Clare (1793-1864)

A heartbreaking and starkly descriptive poem about the cruelty of badger-baiting by the Northamptonshire 'peasant poet', John Clare.

The final verses describe a 'loophole' in the Cruelty to Wild Animals Act of 1835. Whilst it was illegal to allow or encourage dogs to attack a wild badger, there was nothing against keeping a badger as a 'pet'. Live badgers could be dug out and held captive then trotted out for the 'entertainment' of fighting for as long as it could withstand the attacks. A cruel 'sport' indeed.

The badger grunting on his woodland track
With shaggy hide and sharp nose scrowed with black
Roots in the bushes and the woods, and makes
A great high burrow in the ferns and brakes.
With nose on ground he runs an awkward pace,
And anything will beat him in the race.
The shepherd's dog will run him to his den
Followed and hooted by the dogs and men.
The woodman when the hunting comes about
Goes round at night to stop the foxes out
And hurrying through the bushes to the chin
Breaks the old holes, and tumbles headlong in.
When midnight comes a host of dogs and men
Go out and track the badger to his den,
And put a sack within the hole, and lie
Till the old grunting badger passes bye.
He comes and hears—they let the strongest loose.
The old fox hears the noise and drops the goose.
The poacher shoots and hurries from the cry,
And the old hare half wounded buzzes bye.
They get a forked stick to bear him down
And clap the dogs and take him to the town,
And bait him all the day with many dogs,

And laugh and shout and fright the scampering hogs.
He runs along and bites at all he meets:
They shout and hollo down the noisy streets.
He turns about to face the loud uproar
And drives the rebels to their very door.
The frequent stone is hurled where e'er they go;
When badgers fight, then every one's a foe.
The dogs are clapt and urged to join the fray;
The badger turns and drives them all away.
Though scarcely half as big, demure and small,
He fights with dogs for bones and beats them all.
The heavy mastiff, savage in the fray,
Lies down and licks his feet and turns away.
The bulldog knows his match and waxes cold,
The badger grins and never leaves his hold.
He drives the crowd and follows at their heels
And bites them through—the drunkard swears and reels.
The frighted women take the boys away,
The blackguard laughs and hurries on the fray.
He tries to reach the woods, an awkward race,
But sticks and cudgels quickly stop the chase.
He turns again and drives the noisy crowd
And beats the many dogs in noises loud.

He drives away and beats them every one,
And then they loose them all and set them on.
He falls as dead and kicked by boys and men,
Then starts and grins and drives the crowd again;
Till kicked and torn and beaten out he lies
And leaves his hold and cackles, groans, and dies.
Some keep a baited badger tame as hog
And tame him till he follows like the dog.
They urge him on like dogs and show fair play.
He beats and scarcely wounded goes away.
Lapt up as if asleep, he scorns to fly
And seizes any dog that ventures nigh.
Clapt like a dog, he never bites the men
But worries dogs and hurries to his den.
They let him out and turn a harrow down
And there he fights the host of all the town.
He licks the patting hand, and tries to play
And never tries to bite or run away,
And runs away from the noise in hollow trees
Burnt by the boys to get a swarm of bees.

# The Combe by Edward Thomas (1878-1917)

Another poem on the sadness of badger-baiting, this was Edward Thomas's first published piece. A 'combe' is a short valley or hollow on a hillside.

The Combe was ever dark, ancient and dark.
Its mouth is stopped with brambles, thorn, and briar;
And no one scrambles over the sliding chalk
By beech and yew and perishing juniper
Down the half precipices of its sides, with roots
And rabbit holes for steps. The sun of Winter,
The moon of Summer, and all the singing birds
Except the missel-thrush that loves juniper,
Are quite shut out. But far more ancient and dark
The Combe looks since they killed the badger there,
Dug him out and gave him to the hounds,
That most ancient Briton of English beasts.

# The Three Badgers by Lewis Carroll (1832-1892)

In Lewis Carroll's wonderfully whimsical way, he tells the tale of Father Badger and Mother Herring despairing that their children roam and roam and roam. Fortunately, the badger children return, bringing the herring children home with them. Hooray, hooray, hooray!

There be three Badgers on a mossy stone
Beside a dark and covered way:
Each dreams himself a monarch on his throne,
And so they stay and stay -
Though their old Father languishes alone,
They stay, and stay, and stay.

There be three Herrings loitering around,
Longing to share that mossy seat:
Each Herring tries to sing what she has found
That makes Life seem so sweet.
Thus, with a grating and uncertain sound,
They bleat, and bleat, and bleat.

The Mother-Herring, on the salt sea-wave,
Sought vainly for her absent ones:
The Father-Badger, writhing in a cave,
Shrieked out 'Return, my sons!
You shall have buns,' he shrieked, 'if you'll behave!
Yea, buns, and buns, and bun!'

Illustration by Harry Furniss from Lewis Carroll's *Sylvie and Bruno* (1889).

'I fear,' said she, 'your sons have gone astray.
My daughters left me while I slept.'
'Yes'm,' the Badger said: 'it's as you say.
'They should be better kept.'
Thus the poor parents talked the time away,
And wept, and wept, and wept.

'Oh, dear beyond our dearest dreams,
Fairer than all that fairest seems!
To feast the rosy hours away,
To revel in a roundelay!
How blest would be
A life so free -
Ipwergis-Pudding to consume,
And drink the subtle Azzigoom!

'And if, in other days and hours,
Mid other fluffs and other flowers,
The choice were given me how to dine -
"Name what thou wilt: it shall be thine!"
Oh, then I see
The life for me -
Ipwergis-Pudding to consume,
And drink the subtle Azzigoom!'

The Badgers did not care to talk to Fish:
They did not dote on Herrings' songs:
They never had experienced the dish
To which that name belongs:
'And oh, to pinch their tails,' (this was their wish,)
'With tongs, yea, tongs, and tongs!'

'And are not these the Fish,' the Eldest sighed,
'Whose Mother dwells beneath the foam?'
'They are the Fish!' the Second one replied.
'And they have left their home!'
'Oh, wicked Fish,' the Youngest Badger cried,
'To roam, yea, roam, and roam!'

Gently the Badgers trotted to the shore -
The sandy shore that fringed the bay:
Each in his mouth a living Herring bore -
Those aged ones waxed gay:
Clear rang their voices through the ocean's roar,
'Hooray, hooray, hooray!'

Illustration by Harry Furniss from Lewis Carroll's *Sylvie and Bruno* (1889).

An English novelist and poet, Eden Phillpot (1865-1960) was passionate about the area of Dartmoor. As well as being the President of the Dartmoor Preservation Association for several years, he also set many of his novels in the area.

## The Badgers
## Eden Phillpotts

Brocks snuffle from their holt within
A writhen root of black-thorn old,
And moonlight streaks the gashes hold
Of lemon fur from ear to chin,
They stretch and snort and nuff the air,
Then sit, to plan the night's affair.

The neighbours, fox and owl, they heed,
And many whispering scents and sounds
Familiar on the secret rounds,
Then silently make sudden speed,
Paddling away in single file
Adown the eagle fern's dim aisle.

## The Wind in the Willows
## Kenneth Grahame

In Kenneth Grahame's 1908 book *The Wind in the Willows*, Badger is initially introduced as a respected but reclusive sort, living apart from the others in Wild Wood. Rather surprisingly, he turns out to be a stalwart ally and staunch defender of Mr Toad. When things start to go wrong, he helps Toad, Mole and Rat storm Toad Hall and take it back from the weasels and stoats.

Badger from *The Wind in the Willows*.

## The Wicked Tommy Brock

In 1912, Ms Potter released *The Tale of Mr Tod*, a story about two 'disagreeable people', Mr Tod, the villainous fox, and Tommy Brock, the badger, who was a 'short bristly fat waddling person' and 'not nice in his habits'. Ms Potter was, clearly, not a fan of foxes or badgers. In her tale, naughty Tommy Brock tricks old Mr Bouncer – who is a somewhat forgetful babysitter to Benjamin Bunny's young babies, into falling asleep and absconds with the young bunnies. He takes them to Mr Tod's house where he decides to have a nap before preparing the bunnies for tea. A furious Mr Tod discovers the dastardly Tommy Brock in his house, in his bed with his shoes still on, no less! A fight between the two eventually breaks out, giving Benjamin and his cousin Peter Rabbit, an opportunity to whisk the bunnies safely away and back into Mummy Flopsy's arms.

In terms of literature, only Beatrix Potter's badger was a baddie. Most, if not all other classics depicted their badger characters as homely, sensible and loyal.

Above: 'BFFs Forever' duo Rupert Bear and Bill Badger were featured in a comic strip that has been appearing in the *Daily Express* since 1920.

Left: Illustration of Tommy Brock from *The Tale of Mr. Tod*.

Stained glass by Tamsin Abbott.

Left: *Golden Leaves* by Sam Cannon.

# The Sword in the Stone
# T. H. White

A wise and learned badger appears in T. H. White's 1938 publication of *The Sword in the Stone*. His book was subsequently updated in 1958 in the form of *The Once and Future King,* but Wart (the King Arthur to be) and his meeting with Badger remained in the book.

The fabled magician Merlyn transforms Wart into our beloved brock-form and sends him to Badger to learn about his ways, for Merlyn believes badgers to be amongst the wisest of animals.

Badger, whilst seated in his big old reading room, tells Wart a tale as old as time.

The story goes that when God had nearly finished making all of his creatures, He looked at his animal creations and offered them the opportunity for any changes they desired – in hair, teeth or claws – whatever they felt they needed, and each of the animals selected their modifications in turn. But Man chose none, declaring he needed none other than those that God had already chosen for him. God, pleased with Man's ministrations, granted him dominion over all animals. Badger sagely wonders if that 'Order of Dominion' has turned into tyranny.

Original sketch by T. H. White of Badger at home.

# The Return to Narnia C.S. Lewis

The steadfastly loyal badger Trufflehunter features in *Prince Caspian: The Return to Narnia* written by C.S. Lewis and published in 1951. When Prince Caspian is injured fleeing from the usurper Miraz, Trufflehunter takes him back to his den to care for him.

A loyal follower of Aslan and Caspian, the talking badger continues with Caspian on his journey.

# Mr Policeman Badger

Between 1953 and the early 1990s the Royal Society for the Prevention of Accidents (RoSPA) produced books, comics, short tv segments and other merchandise featuring Tufty Fluffytail and his Furryfolk friends, including Mr Policeman Badger, to help children learn the importance of road safety. The characters were created by Elsie Mills MBE as part of her work in child safety. Mr Policeman Badger is kind but stern and tells Willy Weasel quite firmly to 'Never play near the road!'

Badger by British cartoonist Fougasse.

Patrick Chalmers (1872-1942) hailed from Ireland and wrote in a variety of formats on a wide variety of subjects, including hunting, horse racing, cats, dogs and war!

## The Badger
## Patrick R. Chalmers

Last of the night's quaint clan
He goes his way -
A simple gentleman
In sober grey:
To match lone paths of his
In woodlands dim,
The moons of centuries
Have silvered him
Deep in the damp, fresh earth
Roots and rolls,
And builds his winter girth
Of sylvan tolls:
When seek the husbandmen
The Furrow brownm
He hies him to his den
And lays him down.

There may he rest for me,
Nor ever stir
For clamorous obloquy
Of terrier;
Last of the night's quanit clan
He curls in peace -
A friendly gentleman
In grey pelisse!

*A Badger at Avoncliff* by Dru Marland.

*The Last Full Moon* by Hannah Willow.

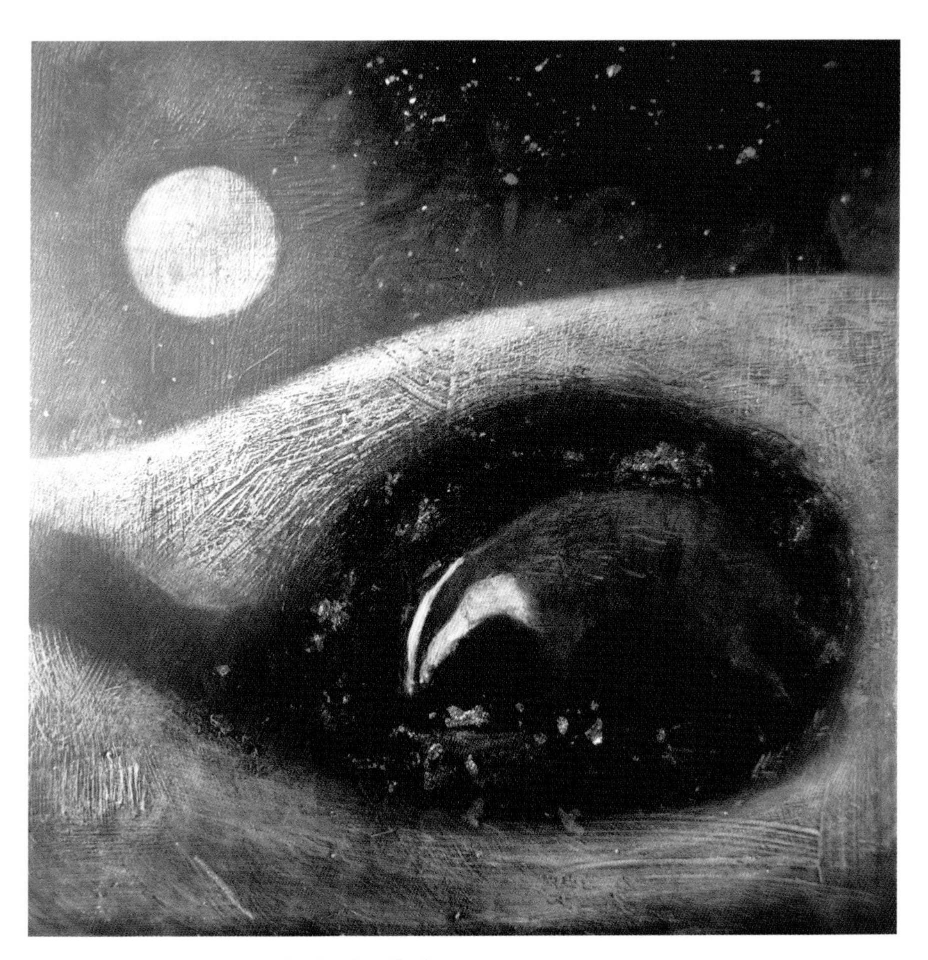

*The Sleeping Earth* by Catherine Hyde.

## Bill Badger

Another 'Bill' Badger was the hero of several books written by Denys Watkins-Pitchford MBE, under the pseudonym 'BB'. Books included *Bill Badger and the Pirates* (1960), in which Bill must defend his barge (*The Wandering Wind*) from Napoleon the cat, his sworn nemesis, and *Bill Badger's Finest Hour* (1961), in which he again, had to defend himself (and Izzybizzy the Hedgehog) from marauding cats – this time, Big Ginger and his gang of robber cats.

## Fantastic Mr Fox's Gentle, 'Respectable' Badger

Unlike Beatrix Potter's feuding fox and badger, Roald Dahl's classic, published in 1970, sees Mr Fox and Badger as firm friends. The book tells the story of Mr Fox trying to outwit the 'wicked' farmers Boggis, Bunce and Bean who, tired of Mr Fox's thieving, plan to starve Mr Fox and his family out. They surround his home and wait, ready for him to risk coming out when he's too hungry. Before long, the siege is affecting not just Mr Fox and his family but the other underground creatures living on the hill.

Mr Fox devises a plan to acquire food for his family and, with a memory of the farms that he claims he would know blindfolded, he begins tunnelling under the ground and, before long, embarks on various raids. He invites Badger and several other burrowing families along to help out.

Soon they tunnel right into Bunce's mighty storehouse and the team load up with goodies.

Felted badger by Karin Celestine.

Badger begins to worry about the stealing and voices his concerns, but Mr Fox heartily defends his actions as that of someone swiping food to feed his hungry children. He accuses Badger of being far too respectable. Badger bristles as he points out there's nothing wrong with being respectable.

Following a good long think about Mr Fox's point about the intent of the horrible farmers to kill them all (and the alternative of starving), Badger continues to help Mr Fox raid all three of the farmers' stores. After an encounter with Rat, some cider and Mrs Bean, they hurry back to Mr Fox's home where Mrs Fox has prepared a fabulous feast with all the loot being sent back along the tunnels. Badger and his family continue to live happily with Mr Fox and the other underground creatures.

In a 2008 film adaptation of the book, Badger turns into the even more respectable Clive Badger, Esq. – a lawyer with the firm Badger, Beaver and Beaver. Also a staunch friend of the wily fox, he was voiced by the actor Bill Murray.

Are stories of badgers and foxes being friends or outright enemies (as in Beatrix Potter's tale) true? Foxes and badgers often forage in the same areas and foxes are even known to take refuge in badger setts, particularly if being hunted. A badger, given its larger size and weight, could seriously injure a fox if necessary and so a fox will preferably avoid confrontational risk. They will largely ignore each other and get on with their own foraging, but fox and badger cubs, however, are known to play and frolic together.

*Beatrix* by Sarah Jones.

More fiercely loyal badgers are to be found in the Redwall series of books, written by Brian Jacques and published between 1986 and 2011. The Badger Lords in the mountain stronghold of Salamandastron are known to be great warriors, while the Badger Mothers are known for their wisdom.

St John's Ambulance encourages 7-10-year-olds to join their Badger Setts and earn badges, learning skills in first aid and helping local communities as well as leadership and communication. Known as 'Badgers', their sweatshirts and badges use a badger logo.

Brockham Badger FC is one of the largest youth football associations in Surrey. Founded in 1979, it now boasts more than 350 boys and girls playing in teams and development squads, playing in, of course, a black and white strip.

The badger's iconic colouring offers a perfect stylised look for badges, logos, and even heraldry. There are beers and balms and everything in between featuring our wonderful brock and its distinctive face.

Even J K Rowlings' Hogwarts School pays homage to the badger for the House of Hufflepuff, which values hard work, loyalty and fair play.

1st
Badger *Meles meles*

*Badger* by Kate Wyatt.

By Of Half Imagined Things.

# Photo Credits and Artworks

**Front cover:** Dod Morrison
**Back cover left to right:**
George Easton, Dod Morrison, Colin Black, Paul Fisher.

**Introduction 4-9**

Page 4, 6: George Easton

Page 8-9: Beverley Thain

**Badger Physiology**

Page 10, 19, 20, 28-29, 34, 42: Dod Morrison

Page 13, 46: Marc Baldwin

Page 14, 27, 35: Paul Fisher

Page 16-17, 33, 36: Colin Black

Page 24, 30, 45, 49: Catriona Komlosi

Page 38-39: Steve Owen

**Badger Watching**

Page 50, 53: Dod Morrison

Page 55, 56: George Easton

Page 57: Wikipedia

Page 58: Beverley Thain

Page 60-61: Colin Black

**Badger Threats**

Page 62: Marc Baldwin

Page 65: Dod Morrison

Page 66: Steve Owen

Page 69, 72-73: Beverley Thain

Page 71: Colin Black

**The Cull**

Page 74, 77, 78, 82: Beverley Thain

Page 79, 81, 84: Steve Owen

Page 85, 88: Catriona Komlosi

Page 87: Colin Black

Page 91: Dod Morrison

Page 93: Paul Fisher

**The Badger Protectors – Wounded Badger Patrol Cheshire**

Page 94, 96-99, 103: Wounded Badger Patrol

Page 102-103: Phil Hatcher

Page 104: Beverley Thain

**Badgers in Myth and Legend**

Page 106, 111, 112, 115, 119, 121, 122, 123: Public domain

**Badgers in Art and Literature**

Page 124, 126: Skyravenwolf, Chris Palmer

Page 132: www.gutenberg.org

Page 135: www.gutenberg.org

Page 136: Public domain

Page 137: Public domain

Page 138: Public domain

Page 139: Photo Jo Byrne

Page 140: Sam Cannon

Page 141: Tamsin Abbott

Page 142: Public domain

Page 144: Public domain

Page 145: Dru Marland

Page 146: Hannah Willow

Page 147: Catherine Hyde

Page 149: Karin Celestine

Page 151: Wikimedia public domain

Page 152: Sarah Jones

Page 153: Logo authorised by Brockham Badgers Chairman, Martin Banks

Page 154: Photo Jo Byrne

Page 155: Kate Wyatt

Page 156-157: Of Half Imagined Things

Page 159: Jane Russ

End papers: linocuts by Jane Russ

Every effort has been made to trace copyright holders of material and acknowledge permission for this publication. The publisher apologises for any errors or omissions to rights holders and would be grateful for notification of credits and corrections that should be included in future reprints or editions of this book.

# Acknowledgements

A giant thank you to Jane Russ for trusting me with another of her projects – especially this one, given I'm a crazy, bonkers badger-fan. Thank you also for your relentless digging to find terrific photos and pictures! It's been fabulous working with you again.

To all the incredibly talented photographers and artists who have gifted us with their wonderful work – HUGE thank you!

To Joana at Graffeg, thank you for pulling all of this series into its magical format.

A massive thank you to Marc Baldwin for his time and effort in reading through the draft and sending through corrections and suggestions and ensuring I "had the badger's head on the right way 'round"… you superstar!

And thank you to my always amazing, forever soulmate, Sean. For loves, laughter and the bestest nature-tastic pal ever…

Sean's suggestion for *The Badger Book*: Take *The Bee Book* and replace every 'bee' with 'badger'.

Me: So… badgers are capable of flapping their wings 230 times a second…?!

The Badger Book
Published in Great Britain in 2021 by Graffeg Limited.



Graffeg Limited, 24 Stradey Park Business Centre, Mwrwg Road, Llangennech, Llanelli, Carmarthenshire, SA14 8YP, Wales, UK. Tel: 01554 824000. www.graffeg.com.

Jo Byrne is hereby identified as the author of this work in accordance with section 77 of the Copyrights, Designs and Patents Act 1988.

A CIP Catalogue record for this book is available from the British Library.



ISBN 9781913634209

1 2 3 4 5 6 7 8 9

# Books in the series

www.graffeg.com

GRAFFEG